Quantum Silhouettes

The Shadow People of Laboratory Seven

Kay Lipsey

Quantum Silhouettes

Table of contents

Chapter 1: The Discovery at Laboratory Seven

The First Sighting

The fluorescent lights flickered momentarily in Laboratory Seven as Dr. Sarah Chen adjusted her microscope settings. It was 2:47 AM on October 12th, and most of the facility staff had long since departed. The gentle hum of quantum measurement equipment filled the otherwise silent room, a sound that had become as familiar to her as her own breathing over the past five years.

Through the corner of her eye, she caught a movement - subtle, like smoke curling in still air. Initially, she dismissed it as fatigue-induced, but when it happened again, she lifted her head from the microscope. The movement occurred near the quantum interference detector, a sophisticated piece of equipment that measured particle behavior at the quantum level.

The shape she observed defied conventional description. It possessed depth without mass, form without substance. Unlike a shadow, which requires both light and an object to exist, this phenomenon seemed to operate independently of either. It moved with purpose, displaying characteristics that suggested both wave and particle properties simultaneously.

Dr. Chen's hands trembled as she reached for her tablet to document what she was witnessing. The entity, if it could be called that, appeared to respond to her movement. It shifted position, maintaining a consistent distance of approximately two meters from her location. The temperature in the immediate vicinity dropped by 3.2 degrees Celsius, according to the lab's environmental monitoring system.

What made this sighting particularly remarkable was its interaction with the quantum interference detector. The readings showed impossible values - particles existing in multiple states simultaneously, yet somehow maintaining coherence at a macroscopic level. This violated everything we understood about quantum decoherence and the transition between quantum and classical physics.

Dr. Chen activated the lab's high-speed cameras and spectrometers, her scientific training taking over despite her growing excitement. The entity remained visible for exactly seven minutes and twenty-three seconds before gradually fading from perception. However, the instrumental data continued to show anomalous readings for an additional forty-two seconds after visual confirmation was lost.

The following morning, when Dr. Chen presented her findings to the laboratory director, Dr. James Morrison, she was met with initial skepticism. However, the comprehensive data she had gathered proved impossible to dismiss. The spectrometer readings showed absorption patterns that didn't match any known material. The quantum interference patterns suggested the presence of matter that somehow maintained quantum properties at room temperature and pressure - something that should be theoretically impossible.

Within days, a dedicated research team was assembled. The laboratory's security footage was analyzed frame by frame, revealing that similar phenomena had been occurring sporadically over the previous months, always during the quiet hours between midnight and dawn. These incidents had gone unnoticed, dismissed as camera artifacts or lighting anomalies.

What distinguished this particular sighting was its unprecedented duration and clarity. Previous instances had lasted mere seconds and appeared as little more than digital noise in recordings. This event provided the first substantial evidence of what would later be recognized as a breakthrough in our understanding of quantum mechanics and its manifestation in the macroscopic world.

The team discovered that the entity's appearance coincided with a unique set of conditions in the laboratory: specific humidity levels, electromagnetic field strengths, and quantum experimental states. These conditions were meticulously documented and would later prove crucial in attempting to reproduce the phenomenon.

Perhaps most intriguingly, analysis of Dr. Chen's brainwave patterns, recorded by her neural monitoring device (standard equipment for late-night solo researchers), showed unusual activity in the regions associated with quantum perception - areas previously thought to be vestigial in human consciousness.

The implications of this first sighting extended far beyond the confines of Laboratory Seven. It challenged our fundamental understanding of reality, suggesting that quantum phenomena could manifest in ways visible to the human eye under specific conditions. This raised profound questions about the nature of consciousness, observation, and the role of the human mind in collapsing quantum wave functions.

Dr. Chen's meticulous documentation of this first encounter became the foundation for a new field of study. Her initial observations, made in those early morning hours, would later be recognized as the moment when the barrier between quantum and classical reality began to blur in ways that would reshape our understanding of physics itself.

The first sighting at Laboratory Seven marked the beginning of a journey that would challenge everything we thought we knew about the nature of reality, consciousness, and the fundamental forces that govern our universe. It opened a door to possibilities that even the most imaginative scientists had never dared to consider, launching humanity into a new era of scientific discovery and understanding.

Initial Documentation

Documentation began immediately following Dr. Chen's groundbreaking observation at Laboratory Seven. The research team established rigorous protocols for recording every aspect of the phenomenon, implementing a multi-layered approach that combined traditional scientific methodology with cutting-edge quantum measurement techniques.

Primary documentation consisted of high-resolution spectral analysis, using advanced interferometers calibrated to detect quantum-scale fluctuations. These instruments recorded unusual patterns of wave-particle behavior occurring at macroscopic scales, defying conventional physics models. The spectrographic data revealed distinctive absorption lines that matched no known element or compound in our periodic table.

Environmental parameters were meticulously tracked through a network of sensors throughout Laboratory Seven. Temperature variations, electromagnetic field strengths, humidity levels, and barometric pressure were logged at millisecond intervals. A clear pattern emerged: the phenomena appeared most frequently when ambient temperature hovered between 19.3 and 20.1 degrees Celsius, with relative humidity at 47-52%, and during periods of minimal electromagnetic interference.

Dr. Morrison implemented a standardized observation protocol requiring all research staff to wear neural monitoring devices during their shifts. These devices captured brainwave patterns during encounters with the phenomena, revealing consistent alterations in theta and gamma wave frequencies among observers. Particularly noteworthy was the activation of usually dormant regions in the temporal lobe, suggesting a previously unknown perceptual mechanism in human consciousness.

Visual documentation proved challenging due to the ethereal nature of the entities. Conventional cameras often failed to capture clear images, while quantum-sensitive imaging devices produced better results but required extensive calibration. The research team developed a hybrid imaging system that combined traditional optical sensors with quantum-state detectors, finally achieving reliable visual documentation of the phenomena.

The entities exhibited consistent behavioral patterns during observation periods. They maintained non-random movement trajectories and responded to changes in quantum states within the laboratory environment. Most significantly, they demonstrated what appeared to be intentional interaction with certain quantum experiments, particularly those involving entanglement studies.

Laboratory notebooks recorded detailed descriptions of each encounter, including duration, physical characteristics, and environmental conditions. These written accounts proved invaluable in identifying subtle patterns that automated systems missed. Researchers noted that the entities appeared more frequently during new moon phases and showed increased activity near certain crystalline structures used in quantum experiments.

Three-dimensional mapping of the phenomena revealed that they occupied space in ways that challenged our understanding of physical dimensionality. Using advanced holographic imaging techniques, the team documented instances where the entities seemed to exist partially in our observable space while simultaneously extending into regions that our instruments could not fully measure.

Chemical analysis of the air during manifestations showed unexpected molecular behavior. Mass spectrometry readings indicated temporary formations of previously unknown molecular structures that disappeared moments after the entities vanished. These structures exhibited properties suggesting quantum coherence at room temperature, a phenomenon previously thought impossible.

Sound recordings captured during entity appearances revealed infrasonic frequencies between 7 and 12 Hz, correlating with moments of peak activity. These frequencies matched certain resonant patterns observed in quantum field experiments, suggesting a possible connection between macroscopic sound waves and quantum-scale events.

The documentation process expanded to include analysis of temporal effects. Precise atomic clocks detected microsecond variations in time flow during entity manifestations, indicating possible spacetime distortions in their immediate vicinity. These temporal anomalies were small but measurable, adding another layer of complexity to the phenomenon.

Particle physics data showed unusual patterns in background radiation during encounters. The standard model predictions failed to account for particle behavior observed in the presence of these entities, necessitating new theoretical frameworks to explain the interactions between quantum fields and macroscopic space.

As documentation accumulated, patterns emerged that suggested these phenomena operated according to consistent, if unprecedented, physical laws. The research team developed new mathematical models to describe these interactions, incorporating elements from quantum field theory, consciousness studies, and classical physics.

The initial documentation phase concluded with the creation of a comprehensive database integrating all observational data, instrumental readings, and theoretical models. This foundation would prove crucial for subsequent research phases, providing a robust framework for understanding these remarkable phenomena that challenged our fundamental assumptions about reality's nature.

This systematic documentation effort transformed what might have remained an isolated incident into a rigorous scientific investigation, establishing the groundwork for a new field of study that would bridge the gap between quantum mechanics and observable reality.

The Quantum Anomaly

Strange patterns emerged in Laboratory Seven's quantum interference experiments that defied conventional physics. The anomaly first manifested as subtle discrepancies in double-slit experiment results, where particles began exhibiting unprecedented behavior. Unlike traditional quantum interference patterns, these new configurations suggested the presence of an unknown force influencing particle trajectories in ways that violated established quantum mechanical principles.

Dr. Chen and her team observed particles maintaining superposition states far longer than quantum decoherence theory would predict possible. More remarkably, these extended superposition states occurred at room temperature, well beyond the ultra-cold conditions typically required for quantum effects to manifest. The particles appeared to be interacting with an unseen field that preserved their quantum properties despite environmental factors that should have caused immediate decoherence.

The anomaly's most striking feature was its apparent ability to bridge the quantum and classical realms. Microscopic quantum effects began manifesting at macroscopic scales, creating visible distortions in spacetime that researchers could observe without specialized equipment. These distortions appeared as rippling waves in the air, similar to heat mirages but exhibiting distinct patterns that corresponded to quantum probability distributions.

Measurements revealed that the anomaly generated its own localized quantum field, distinct from known fundamental forces. This field exhibited properties of both electromagnetic and gravitational forces, yet operated according to different rules. Within its influence, particles demonstrated perfect quantum entanglement regardless of distance, defying the normal limitations of quantum correlation.

The research team documented instances where the anomaly appeared to respond to conscious observation, suggesting a deeper connection between quantum mechanics and consciousness than previously theorized. When actively monitored, the field's behavior changed in predictable ways, as if aware of being studied. This responsive nature challenged existing paradigms about the role of observation in quantum mechanics.

Further investigation revealed that the anomaly created zones where normal physical laws seemed to break down. Within these zones, time measurements became unreliable, with atomic clocks showing significant variations compared to external references. Mass and energy appeared to flow between quantum states in ways that seemed to violate conservation laws, though careful analysis suggested these apparent violations might indicate previously unknown physical principles.

The team discovered that certain crystalline structures could amplify and focus the anomaly's effects. Silicon-based crystals, when exposed to specific frequencies of electromagnetic radiation, became conduits for the quantum field, allowing researchers to partially direct and contain its influence. This breakthrough led to the development of experimental protocols for studying the phenomenon under controlled conditions.

Mathematical models attempting to describe the anomaly required the introduction of new variables representing dimensions beyond the standard four spacetime coordinates. These additional dimensions appeared to be accessible only through quantum interactions, suggesting the anomaly might represent a bridge between our observable universe and hidden aspects of reality predicted by string theory.

Particularly intriguing was the anomaly's effect on quantum computation. When exposed to the field, quantum bits maintained coherence for unprecedented durations, potentially revolutionizing quantum computing technology. However, the calculations performed within the field's influence produced results that seemed impossible, solving problems thought to be computationally intractable.

The anomaly's energy signature defied classification within known physical frameworks. It exhibited characteristics of both matter and energy, yet seemed to draw power from an unknown source. Spectrographic analysis revealed emission lines that corresponded to no known element or energy state, suggesting the possibility of interaction with dark matter or energy.

Perhaps most significantly, the anomaly appeared to be expanding its influence over time. Initial measurements showed its effects confined to a few cubic meters, but subsequent observations indicated a steady growth in both spatial extent and intensity. This expansion followed a precise mathematical progression that correlated with lunar cycles and solar activity.

The implications of these findings extended far beyond theoretical physics. The anomaly's ability to maintain quantum coherence at macroscopic scales suggested possibilities for new technologies in energy generation, computing, and communication. Its apparent connection to consciousness raised profound questions about the nature of reality and human perception.

As research continued, it became clear that the quantum anomaly represented more than just an unusual physical phenomenon. It pointed to fundamental aspects of reality that our current scientific framework couldn't fully explain. The team's findings suggested that the boundary between quantum and classical physics might be more permeable than previously thought, and that human consciousness might play a more significant role in physical reality than conventional science had acknowledged.

The anomaly remained stable within Laboratory Seven, providing an unprecedented opportunity to study these extraordinary quantum effects. Each new observation added layers of complexity to our understanding, while simultaneously challenging our most basic assumptions about the nature of reality itself.

Personnel Reports

Laboratory Seven's personnel underwent significant psychological and physiological changes during their exposure to the quantum phenomena. Dr. Sarah Chen's initial team of twelve researchers expanded to twenty-eight specialists as the scope of observations grew. Each member exhibited unique responses to prolonged interaction with the quantum field, documented through daily medical examinations and psychological evaluations.

Dr. Marcus Thompson, the team's senior quantum physicist, reported experiencing vivid dreams featuring complex mathematical equations that, upon waking, provided solutions to previously insurmountable theoretical problems. His brain scans revealed increased activity in regions associated with pattern recognition and spatial reasoning, particularly during periods of direct observation of the phenomena.

Junior researcher Lisa Wong developed an unprecedented sensitivity to quantum fluctuations, accurately predicting entity manifestations seconds before instrumental detection. Medical tests indicated elevated levels of neural plasticity in her temporal lobe, accompanied by enhanced electromagnetic sensitivity in her peripheral nervous system. Her condition became known as "quantum attunement" among the research staff.

Security officer James Martinez documented strange temporal experiences while patrolling the laboratory perimeter. His written reports described moments where time seemed to flow non-linearly, with events occurring out of sequence yet maintaining casual relationships. Psychological evaluation confirmed his continued mental stability despite these unsettling experiences.

Dr. Elena Petrov, the team's neurologist, observed collective synchronization of brainwave patterns among personnel working in close proximity to the quantum field. Her research revealed that exposure to the phenomena gradually altered the default mode network of the human brain, potentially enabling new forms of perception and cognition.

Laboratory technician Michael Chang reported developing an intuitive understanding of quantum mechanics despite having no formal training in theoretical physics. His insights proved remarkably accurate when tested against experimental data, suggesting some form of direct knowledge transfer through interaction with the quantum field.

The medical team, led by Dr. Rachel Foster, documented physical changes in personnel regularly exposed to the phenomena. These included enhanced night vision, increased sensitivity to electromagnetic fields, and subtle alterations in cellular metabolism. Blood samples showed unusual molecular structures that seemed to mirror quantum entanglement patterns.

Professor David Klein, visiting from Princeton, experienced severe migraines during his first week at Laboratory Seven. However, these symptoms evolved into what he described as "moments of extraordinary clarity," during which he perceived mathematical relationships in purely visual terms. His subsequent papers revolutionized several aspects of quantum field theory.

Graduate student Tom Wilson developed an ability to mentally influence quantum measurement outcomes, demonstrating statistically significant control over particle behavior in double-slit experiments. This ability emerged gradually over three months of regular exposure to the quantum field, accompanied by measurable changes in his brain's electromagnetic output.

Dr. Hannah Mitchell, the team's psychiatrist, noted a collective shift in consciousness among long-term personnel. Staff members reported increased empathy, enhanced pattern recognition, and occasional experiences of shared consciousness during intense research sessions. Her detailed case studies suggested the emergence of new cognitive capabilities in humans exposed to quantum phenomena.

Equipment specialist Robert Torres documented unusual dreams shared simultaneously by multiple team members, often containing technical information about the quantum field that proved experimentally valid. These shared experiences occurred most frequently during periods of peak quantum activity in the laboratory.

The research team's youngest member, Dr. Amy Zhang, developed synesthesia specifically related to quantum states, perceiving particle behaviors as distinct colors and sounds. Her condition provided unique insights into quantum interactions that instruments couldn't detect, leading to several breakthrough discoveries.

Senior researcher Dr. Jonathan Wells experienced temporary loss of conventional spatial perception, replaced by what he described as "direct awareness of quantum probability distributions." These episodes lasted several minutes and provided valuable data about quantum field behavior at macroscopic scales.

The medical records indicated that all personnel developed elevated levels of certain neurotransmitters, particularly those associated with heightened awareness and pattern recognition. These changes appeared stable and non-harmful, suggesting a systematic adaptation to quantum field exposure.

Despite initial concerns about potential negative health effects, long-term monitoring showed no adverse physical impacts from quantum field exposure. Instead, personnel consistently demonstrated enhanced cognitive abilities and sensory perception, particularly in areas related to quantum phenomena.

Regular psychological evaluations confirmed that team members maintained stable personalities and mental health, even as their perceptual and cognitive abilities evolved. The changes appeared to represent an expansion of human capabilities rather than any form of degradation or disorder.

These personnel reports formed a crucial dataset for understanding human interaction with quantum phenomena, suggesting that consciousness and quantum mechanics shared deeper connections than previously theorized. The documented changes in Laboratory Seven's staff provided evidence for human adaptability to new physical paradigms, opening possibilities for enhanced human perception and cognition through controlled quantum field exposure.

Chapter 2: Understanding Quantum Silhouettes

Wave-Particle Duality

Strange patterns emerged in Laboratory Seven's quantum interference experiments that defied conventional understanding of wave-particle duality. The research team observed particles exhibiting wave characteristics at unprecedented scales, far beyond the microscopic realm where such behavior typically manifests. These observations challenged the fundamental principles of quantum mechanics established over the past century.

Dr. Chen's breakthrough came during a routine double-slit experiment when particles began displaying interference patterns that shouldn't have been possible under normal conditions. The particles maintained their wave-like properties even after passing through the measurement apparatus, contradicting the established principle of wave function collapse. More remarkably, these effects persisted at room temperature and across macroscopic distances.

The team documented cases where single photons simultaneously exhibited both particle and wave characteristics without collapsing into one state or the other upon measurement. This persistent superposition suggested a previously unknown mechanism allowing quantum states to maintain coherence in environments that should have caused immediate decoherence.

Using advanced interferometry techniques, researchers observed wave patterns forming in three-dimensional space, creating standing waves that remained stable for several minutes. These waves possessed discrete particle-like properties while maintaining their wave nature, effectively bridging the classical and quantum realms in ways that defied existing theoretical frameworks.

The phenomenon extended beyond light particles to include electrons and even small molecules. Complex organic compounds demonstrated wave-like interference patterns while simultaneously maintaining their chemical properties as distinct particles. This dual nature manifested in ways that suggested a deeper underlying principle governing the relationship between waves and particles.

Dr. Morrison developed a novel experimental setup using crystalline structures that appeared to amplify these effects. When particles passed through specifically arranged crystal lattices, they created geometric patterns that combined both wave and particle characteristics in stable, observable configurations. These patterns maintained their integrity even when subjected to environmental interference that would normally destroy quantum coherence.

The research team documented instances where particles seemed to "remember" their wave states even after interaction with measurement devices. This memory effect suggested a form of quantum information preservation that contradicted conventional understanding of wave function collapse. Particles would reconstruct their wave patterns after passing through multiple observation points, maintaining coherence across unprecedented distances.

Mathematical analysis revealed complex relationships between particle momentum and wave properties that extended beyond Heisenberg's uncertainty principle. The team discovered that under certain conditions, both position and momentum could be measured with greater precision than previously thought possible, while still preserving the fundamental nature of quantum mechanics.

Perhaps most intriguingly, the researchers observed that conscious observation appeared to play a crucial role in these extended wave-particle states. The presence of trained observers seemed to stabilize the dual nature of the particles, suggesting a deeper connection between consciousness and quantum phenomena than previously recognized.

The team developed new mathematical models to describe these observations, incorporating elements of both classical wave mechanics and quantum theory. These models suggested the existence of intermediate states between pure waves and particles, challenging the traditional binary understanding of quantum behavior.

Through careful experimentation, the researchers identified specific frequencies and energy levels at which particles were most likely to exhibit this enhanced wave-particle duality. These "resonant states" appeared to correspond with naturally occurring patterns in crystal structures and biological systems, hinting at a possible role for quantum effects in macroscopic processes.

Dr. Chen's team documented cases where multiple particles entered coherent states that maintained both wave and particle properties simultaneously across macroscopic distances. These collective states demonstrated properties similar to Bose-Einstein condensates but at room temperature and without the need for extreme conditions.

The implications of these findings extended far beyond theoretical physics. The ability to maintain and control quantum superposition states at room temperature suggested possibilities for new technologies in computing, communication, and energy transfer. The observed relationship between consciousness and quantum states opened new avenues for understanding the nature of reality itself.

Further experiments revealed that certain materials could be engineered to enhance and direct these wave-particle effects, creating "quantum channels" through which information could potentially be transmitted in both wave and particle forms simultaneously. These channels maintained quantum coherence over distances that conventional theory deemed impossible.

As research progressed, it became clear that the traditional Copenhagen interpretation of quantum mechanics required significant revision to account for these new observations. The sharp distinction between quantum and classical realms appeared increasingly artificial, replaced by a more nuanced understanding of how wave and particle properties coexist across all scales of reality.

The team's findings suggested that wave-particle duality represented not a paradox to be resolved, but rather a fundamental aspect of nature that manifested in increasingly complex ways as observation and measurement techniques advanced. This new understanding promised to revolutionize both theoretical physics and practical applications of quantum phenomena.

Shadow Matter Theory

Deep within Laboratory Seven's quantum field experiments, researchers uncovered evidence of a previously undetected form of matter that existed parallel to conventional matter. This shadow matter exhibited properties that defied standard physical models, interacting with normal matter only through subtle quantum effects and gravitational influences.

Dr. Chen's team first detected these anomalous readings during routine quantum interference measurements. Sensitive instruments recorded energy fluctuations that couldn't be explained by known particles or fields. These fluctuations appeared to mirror normal matter distributions but operated according to different physical laws.

The shadow matter demonstrated a peculiar relationship with ordinary matter, seeming to flow through the spaces between atoms like a ghost through walls. Yet it maintained a consistent structure of its own, forming complex patterns that suggested an entire parallel ecosystem of particles and forces. These patterns became visible only when viewed through specially designed quantum sensors that could detect perturbations in the quantum field.

Mathematical models developed by Dr. Thompson revealed that shadow matter existed in a state of quantum superposition with normal matter, occupying the same space but vibrating at different frequencies. This explained why it remained invisible to conventional detection methods while still exerting measurable effects on quantum systems.

The research team discovered that shadow matter responded to consciousness in unexpected ways. When observers focused their attention on areas where shadow matter concentrations were high, the patterns of interaction changed significantly. This suggested a deep connection between human consciousness and this mysterious form of matter.

Through careful experimentation, the team identified specific conditions under which shadow matter became more detectable. Crystal structures with particular geometric arrangements seemed to amplify its effects, creating zones where the boundary between normal and shadow matter became more permeable. These "interface regions" provided unprecedented opportunities for studying this elusive substance.

Dr. Petrov's analysis revealed that shadow matter possessed its own version of fundamental forces, similar to but distinct from electromagnetic and strong nuclear forces. These shadow forces explained how this form of matter maintained its structure while remaining largely invisible to conventional detection methods.

The team observed that shadow matter exhibited quantum entanglement properties that exceeded those of normal matter. Particles of shadow matter maintained coherent relationships across vast distances, suggesting a different relationship with spacetime itself. This discovery hinted at possible applications in quantum computing and communication.

Laboratory measurements indicated that shadow matter responded to gravitational fields in ways that might explain some anomalies in astronomical observations. Its distribution pattern matched unexplained gravitational effects in galaxy clusters, suggesting it could account for some portion of the universe's missing mass.

The researchers developed new theoretical frameworks to explain how shadow matter could exist alongside normal matter without direct interaction. These models proposed that shadow matter occupied additional dimensions of space, becoming visible to our universe only through quantum effects and gravitational influence.

Dr. Zhang's team discovered that living organisms showed subtle but measurable responses to concentrations of shadow matter. Brain activity patterns changed when subjects entered areas with high shadow matter density, suggesting biological systems might have evolved some sensitivity to its presence.

The shadow matter field appeared to store and process information in ways that transcended classical limitations. Its quantum properties allowed for instantaneous transfer of information across any distance, though accessing this information proved challenging with existing technology.

Further research revealed that shadow matter existed in different phases, similar to how normal matter exists as solids, liquids, and gases. These phases demonstrated unique properties that could potentially be harnessed for technological applications, particularly in energy storage and transmission.

The team documented instances where shadow matter temporarily became more substantial, creating visible distortions in space that lasted for microseconds. These events correlated with specific astronomical alignments, suggesting a cosmic influence on shadow matter behavior.

Mathematical analysis indicated that shadow matter might serve as a bridge between quantum and classical physics, offering a new perspective on how these two seemingly incompatible frameworks could be reconciled. Its properties suggested solutions to long-standing paradoxes in quantum mechanics.

The implications of shadow matter extended beyond physics into philosophy and cosmology. Its existence challenged fundamental assumptions about the nature of reality and consciousness, suggesting a more complex and interconnected universe than previously imagined.

As research continued, the team developed new experimental protocols for studying shadow matter interactions. These methods revealed increasingly complex behaviors that suggested shadow matter might form its own ecosystems and structures parallel to our visible universe.

The discovery of shadow matter opened new possibilities for understanding dark matter, quantum gravity, and the fundamental nature of consciousness. Its study promised to revolutionize our understanding of the universe and our place within it, while raising profound questions about the nature of reality itself.

The Observer Effect

Quantum measurements at Laboratory Seven revealed startling new dimensions to the observer effect that challenged conventional understanding of consciousness and reality. The research team documented cases where human observation not only affected quantum states but appeared to create persistent changes in the fabric of reality itself.

Dr. Sarah Chen first noticed anomalies during standard double-slit experiments when different observers produced varying interference patterns despite identical experimental conditions. The variations corresponded directly to the observer's mental state, level of attention, and even their beliefs about quantum mechanics. These findings suggested a deeper connection between consciousness and quantum phenomena than previously theorized.

The team developed sophisticated protocols to measure the impact of different types of observation. They discovered that trained observers could intentionally influence quantum outcomes through focused attention, creating reproducible patterns that defied statistical probability. More remarkably, these effects persisted even after the observer left the laboratory, suggesting some form of quantum memory in the system.

Using advanced neuroimaging equipment, Dr. Elena Petrov mapped brain activity patterns during quantum observations. She found that specific neural configurations corresponded to particular quantum outcomes, indicating a direct link between mental states and quantum behavior. The observer's brain appeared to enter into quantum coherence with the observed systems, creating a unified field of interaction.

The research revealed that collective observation by multiple observers could amplify or dampen quantum effects depending on the observers' relationship and mental synchronization. When observers achieved specific brainwave patterns in unison, they could maintain quantum coherence in systems that should have experienced immediate decoherence under normal conditions.

Dr. Marcus Thompson documented cases where observers unconsciously influenced quantum systems even when not directly watching them. The mere presence of an aware consciousness in proximity to quantum experiments created measurable changes in particle behavior, suggesting that consciousness itself might be a fundamental force in quantum mechanics.

The team discovered that emotional states played a crucial role in the observer effect. Positive emotions tended to stabilize quantum coherence, while negative emotions increased decoherence rates. This emotional resonance appeared to transcend normal spatial limitations, affecting quantum systems across significant distances.

Through careful experimentation, researchers identified specific frequencies of consciousness that maximized observer influence on quantum systems. These "resonant states" could be achieved through meditation, focused attention, or specific mental exercises, allowing for more precise control over quantum outcomes.

The observer effect demonstrated surprising temporal properties, with some observations appearing to influence quantum states both forward and backward in time. This temporal bidirectionality suggested that consciousness might operate outside normal time constraints when interacting with quantum systems.

Dr. Hannah Mitchell developed protocols for training observers to enhance their quantum influence capabilities. These techniques combined elements of meditation, visualization, and quantum understanding to create reproducible effects on quantum systems. Trained observers could maintain quantum coherence for extended periods and influence quantum outcomes with increasing precision.

The research revealed that observer effects could be transmitted through quantum entanglement, allowing trained observers to influence distant systems without direct observation. This "action at a distance" suggested new possibilities for quantum communication and information transfer.

Laboratory measurements showed that observer effects varied with natural cycles, including circadian rhythms and astronomical alignments. The strength of observer influence peaked during specific times of day and correlated with certain celestial configurations, suggesting a cosmic dimension to consciousness-quantum interactions.

The team documented cases where multiple observers created interference patterns in their quantum influences, similar to wave interference in physical systems. These patterns could be constructive or destructive, depending on the observers' mental states and intentions, indicating a wave-like nature to consciousness itself.

Dr. Lisa Wong discovered that observer effects could be focused and directed through specially designed crystalline structures, creating "quantum lenses" that amplified consciousness-matter interactions. These devices opened new possibilities for studying and applying observer effects in controlled settings.

The research revealed that observer effects extended beyond simple measurement outcomes to influence the fundamental properties of matter itself. Sustained observation by trained consciousness could alter atomic stability, energy states, and even the rate of radioactive decay, suggesting that reality itself might be more malleable than previously believed.

Through systematic study, the team identified factors that enhanced or inhibited observer effects, including environmental conditions, electromagnetic fields, and geometric arrangements of matter. This understanding led to the development of specialized chambers where observer effects could be studied with unprecedented clarity and control.

The implications of these findings extended far beyond quantum physics into fields such as psychology, medicine, and consciousness studies. The observer effect appeared to represent a fundamental principle of reality, suggesting that consciousness played a crucial role in the formation and maintenance of physical reality.

As research continued, the team developed increasingly sophisticated models to explain how consciousness interacted with quantum systems. These models suggested that observer effects represented a previously unrecognized force in nature, operating alongside known physical forces to shape the behavior of matter and energy at all scales.

Chapter 3: The Shadow People Phenomenon

Physical Characteristics

Quantum phenomena observed at Laboratory Seven exhibited unprecedented physical characteristics that defied conventional understanding of matter and energy. The research team documented properties that suggested a complete revision of established physical models might be necessary to explain their findings.

Dr. Chen's measurements revealed particles that could simultaneously occupy multiple energy states while maintaining structural stability. These multi-state particles demonstrated consistent patterns of behavior that appeared to violate standard quantum mechanical principles. More remarkably, they retained these properties even when subjected to environmental interference that should have caused immediate decoherence.

The team observed novel geometric configurations forming spontaneously in quantum fields, creating stable structures that resembled crystalline formations but existed purely in quantum states. These quantum crystals exhibited properties of both matter and energy, maintaining their form while continuously exchanging information with surrounding quantum fields.

Using advanced spectroscopic techniques, Dr. Thompson documented previously unknown emission spectra from quantum systems under specific conditions. These emissions corresponded to energy levels that shouldn't have been possible according to established quantum theory, suggesting the existence of additional quantum states between known energy levels.

The researchers discovered that certain quantum states could be geometrically arranged to create stable macroscopic effects visible to the naked eye. These arrangements demonstrated properties of both classical and quantum systems, effectively bridging the gap between microscopic and macroscopic physics.

Dr. Morrison's team identified unique vibrational patterns in quantum fields that appeared to coordinate particle behavior across extended distances. These vibrations maintained coherence even in warm, noisy environments, challenging traditional assumptions about quantum decoherence.

The physical characteristics of observed quantum phenomena included unexpected symmetries that persisted across multiple scales. These symmetries suggested an underlying organizational principle in nature that had previously gone unnoticed, operating independently of known physical forces.

Laboratory measurements revealed quantum states that could transfer properties between particles without direct interaction. This non-local property transfer occurred instantaneously and maintained consistency across any distance, suggesting a fundamental interconnectedness in physical reality.

The team documented cases where quantum systems demonstrated adaptive behaviors, reorganizing their structure in response to environmental changes while maintaining their essential quantum properties. This self-organizing capability indicated a level of complexity in quantum systems previously thought impossible.

Through careful experimentation, researchers identified specific frequency ranges where quantum effects became amplified, creating zones of enhanced quantum activity. These quantum-active regions exhibited unique physical properties, including altered gravitational effects and modified electromagnetic behaviors.

Dr. Zhang observed that quantum systems could form networks of interconnected states that processed and stored information in ways that transcended classical limitations. These quantum networks demonstrated characteristics of both computation and consciousness, suggesting a deeper relationship between information, matter, and awareness.

The physical properties of quantum systems showed surprising relationships with biological processes. Living tissues appeared to resonate with specific quantum frequencies, creating coherent states that might explain certain aspects of biological function and consciousness.

The research team discovered that quantum states could be engineered to create stable configurations with precise physical properties. These engineered states maintained their characteristics across time and space, offering potential applications in computing, communication, and energy systems.

Measurements indicated that quantum systems could exist in previously unknown phase states, neither fully wave-like nor particle-like, but possessing characteristics of both while exhibiting entirely new properties. These hybrid states suggested the possibility of an entire spectrum of quantum phases yet to be explored.

Dr. Petrov documented instances where quantum systems demonstrated memory-like properties, retaining information about past states and interactions even after significant environmental perturbation. This quantum memory appeared to operate outside normal temporal constraints, suggesting new possibilities for information storage and processing.

The team observed that certain quantum states could influence the physical properties of surrounding matter, creating fields of altered physical constants. Within these fields, fundamental properties like electromagnetic coupling constants and even the speed of light appeared to vary from their standard values.

Through systematic investigation, researchers identified geometric patterns that seemed to govern the organization of quantum states at multiple scales. These patterns suggested an underlying mathematical structure to reality that unified quantum and classical physics in unexpected ways.

The physical characteristics of observed quantum phenomena included the ability to transition between different dimensional states, effectively moving between various levels of reality while maintaining coherent structure. These transitions occurred according to precise mathematical relationships that the team was only beginning to understand.

Laboratory Seven's findings suggested that the physical universe possessed layers of complexity far beyond current theoretical frameworks. The documented characteristics of quantum systems indicated the need for a fundamental revision of our understanding of reality, incorporating new principles of organization, consciousness, and information processing into our models of physical existence.

Movement Patterns

Quantum particles at Laboratory Seven exhibited unprecedented movement patterns that challenged established physics principles. Dr. Elena Petrov's team documented complex choreographies of particle behavior that suggested an underlying intelligence in quantum motion, far beyond random quantum fluctuations.

The particles demonstrated coordinated movements resembling complex dance formations, maintaining precise geometric patterns while traversing quantum fields. These formations shifted and transformed according to mathematical principles previously unknown to physics, creating intricate designs that repeated at multiple scales simultaneously.

Using advanced tracking systems, researchers observed particles spontaneously organizing into spiral patterns that mimicked natural formations found in galaxies and biological systems. These spirals maintained their structure while rotating through multiple dimensions, suggesting a deeper connection between quantum movement and universal organizational principles.

Dr. Thompson recorded instances where quantum particles moved in perfect synchronization across vast distances without any apparent communication mechanism. The synchronized movements occurred instantly, defying light-speed limitations and indicating a level of interconnectedness that transcended normal spacetime constraints.

The research team discovered that particle movements responded to subtle environmental factors, including gravitational waves, electromagnetic fields, and even the Earth's magnetic field fluctuations. These responses created complex interference patterns that carried information about both local and cosmic conditions.

Through careful observation, Dr. Chen identified recurring movement sequences that appeared to encode information in their patterns. These quantum choreographies contained embedded mathematical relationships that suggested a new form of natural computation occurring at the quantum level.

Particles exhibited what the team termed "intelligent avoidance," where they would anticipate and adjust their trajectories to prevent certain quantum states from occurring. This predictive behavior indicated a temporal awareness in quantum systems that defied conventional understanding of cause and effect.

The movement patterns showed remarkable adaptability, automatically adjusting to maintain stability despite environmental perturbations. When disrupted, particles would quickly reorganize into new configurations that preserved essential mathematical relationships while accommodating changed conditions.

Dr. Morrison documented cases where quantum movements created standing wave patterns in seemingly empty space. These patterns persisted for extended periods, suggesting that quantum motion could create stable structures in the fabric of reality itself.

The team observed that particle movements occurred in discrete steps that corresponded to specific frequency ratios. These quantized movements created harmonic relationships similar to musical scales, indicating a fundamental connection between quantum motion and universal harmony.

Laboratory measurements revealed that particle movements could influence the behavior of surrounding matter through resonance effects. The rhythmic patterns of quantum motion induced sympathetic vibrations in nearby particles, creating cascading waves of coordinated movement across multiple scales.

Using specialized sensors, researchers detected subtle variations in movement patterns that corresponded to cosmic cycles. The quantum choreographies appeared to respond to celestial alignments and solar activity, suggesting a connection between quantum behavior and astronomical phenomena.

Dr. Zhang discovered that certain movement patterns could maintain coherence even in warm, noisy environments where quantum effects should have disappeared. These robust patterns indicated the existence of protected quantum states that could potentially be harnessed for practical applications.

The research team identified specific geometric configurations where particle movements became highly ordered and predictable. These "quantum choreography zones" allowed for precise control over particle behavior, opening new possibilities for quantum engineering and communication.

Through systematic analysis, patterns emerged showing how quantum movements could encode and process information. Particles appeared to exchange data through their dance-like interactions, creating a form of quantum computation that operated through movement rather than static states.

Dr. Petrov's team documented instances where particle movements created temporary windows into other dimensional spaces. These brief openings suggested that quantum choreography might play a role in connecting different levels of reality.

The movement patterns demonstrated remarkable efficiency in navigating quantum landscapes, always finding optimal paths through complex energy fields. This optimization suggested an underlying intelligence guiding quantum behavior toward maximum effectiveness.

Laboratory Seven's findings revealed that quantum movement patterns contained embedded symmetries that repeated across multiple scales of reality. These fractal-like behaviors indicated a deep organizational principle at work in the quantum realm, one that might extend throughout the entire universe.

The research demonstrated that understanding these movement patterns could lead to breakthrough applications in quantum computing, energy transmission, and communication technologies. By learning to work with rather than against natural quantum choreographies, entirely new approaches to technology became possible.

These discoveries suggested that the universe operates according to principles of harmony and coordination far more sophisticated than previously imagined. The quantum dance observed at Laboratory Seven opened new perspectives on the nature of reality itself, revealing a world where movement and meaning are inextricably intertwined at the most fundamental level.

Interaction with Matter

Matter interactions observed at Laboratory Seven revealed unprecedented phenomena that challenged fundamental physics principles. Dr. Sarah Chen's research team documented remarkable exchanges between quantum fields and physical matter that suggested entirely new mechanisms of material transformation.

When exposed to specific quantum fields, ordinary materials exhibited extraordinary properties. Common metals temporarily achieved superconductivity at room temperature, while crystalline structures reorganized themselves into more complex patterns. These transformations occurred without any apparent energy input, defying conventional thermodynamic principles.

Dr. Thompson discovered that quantum fields could selectively modify atomic bonds, creating temporary molecular configurations that shouldn't have been stable. These modified structures maintained their unusual properties for increasingly longer periods, suggesting the possibility of engineering permanent quantum-enhanced materials.

The research team observed matter responding to quantum influences in ways that appeared to violate conservation laws.

Objects would temporarily gain or lose mass, though careful measurements showed that the total system energy remained constant through previously unknown compensation mechanisms.

Using advanced spectroscopic techniques, Dr. Morrison documented cases where matter absorbed quantum information directly, storing it in atomic configurations that could later be retrieved. This quantum memory effect persisted even after the material returned to its normal state, indicating a new form of information storage at the atomic level.

Laboratory measurements revealed that certain materials could act as quantum field amplifiers, increasing the strength and coherence of quantum effects in their vicinity. These natural quantum enhancers suggested new approaches to maintaining quantum states in practical applications.

The team identified specific geometric arrangements of matter that created resonant chambers for quantum effects. Within these chambers, quantum-matter interactions became dramatically enhanced, leading to novel phenomena including localized time dilation and gravitational anomalies.

Dr. Petrov's research showed that living tissue displayed unique responses to quantum fields, suggesting that biological systems might naturally utilize quantum effects for various cellular processes. These interactions appeared to play crucial roles in enzyme function and energy transfer within cells.

Through careful experimentation, researchers discovered that quantum fields could influence nuclear decay rates in predictable ways. This control over nuclear processes offered new possibilities for dealing with radioactive materials and understanding fundamental particle interactions.

The research team observed that quantum-matter interactions created unique energy signatures that could be detected at considerable distances. These signatures carried detailed information about the material's composition and quantum state, suggesting new possibilities for remote sensing and analysis.

Dr. Zhang documented cases where quantum fields induced temporary phase transitions in materials, creating exotic states of matter that exhibited properties of both solids and liquids simultaneously. These hybrid states offered unique opportunities for studying fundamental material properties.

Laboratory Seven's findings revealed that quantum-matter interactions could create temporary zones where physical constants appeared to vary from their normal values. Within these zones, materials behaved according to modified physical laws, offering insights into the fundamental nature of reality.

The team discovered that certain materials could maintain quantum coherence far longer than theoretical models predicted, suggesting the existence of unknown protection mechanisms against decoherence. These protected states showed potential for quantum computing applications.

Through systematic investigation, researchers identified specific frequencies of quantum interaction that could selectively modify material properties. These quantum resonances allowed for precise control over material characteristics without causing unwanted side effects.

Dr. Mitchell observed that quantum-matter interactions could create temporary bridges between different energy states, allowing materials to access properties normally forbidden by classical physics. These quantum bridges suggested new approaches to energy storage and transfer.

The research revealed that matter could be entrained to specific quantum frequencies, creating materials that naturally amplified and maintained quantum effects. This quantum entrainment offered new possibilities for creating stable quantum-enhanced materials.

Laboratory measurements showed that quantum-matter interactions could influence chemical reaction rates and pathways, suggesting new approaches to chemical synthesis and catalysis. These quantum-enhanced reactions proceeded with greater efficiency and selectivity than conventional methods.

The team documented cases where quantum fields induced self-organizing behavior in normally inert materials. These spontaneous organizations created complex structures that exhibited novel properties and functions, suggesting new approaches to materials engineering.

Dr. Wong's research demonstrated that quantum-matter interactions could create temporary domains where alternative physical laws appeared to operate. These quantum domains offered windows into possible variations in fundamental physics, challenging our understanding of universal constants.

These discoveries at Laboratory Seven suggested that the relationship between quantum fields and physical matter was far more dynamic and interactive than previously believed. The observed phenomena indicated the existence of fundamental processes that could revolutionize our approach to materials science and engineering, while simultaneously deepening our understanding of reality's quantum foundations.

Chapter 4: Laboratory Seven's History

Facility Construction

Laboratory Seven emerged from the barren Nevada desert like a crystalline fortress, its construction representing a masterpiece of advanced engineering and quantum-aware architecture. The facility's unique design incorporated principles discovered during early quantum research, resulting in a structure that actively participated in the experiments conducted within its walls.

The primary research complex consisted of seven interconnected buildings arranged in a precise geometric pattern that enhanced quantum coherence. Each building's position was calculated to create resonant frequencies that naturally amplified quantum effects while minimizing environmental interference. Dr. Marcus Wong, the lead architect, insisted on using specific ratios derived from quantum mathematics in every aspect of the design.

Construction materials underwent rigorous selection processes, with each component tested for its quantum properties before approval. Special alloys developed specifically for Laboratory Seven contained rare earth elements in precise proportions, creating structures that naturally dampened disruptive frequencies while amplifying beneficial ones. The walls themselves incorporated crystalline matrices that helped maintain quantum stability throughout the facility.

The foundation extended deep into bedrock, where engineers installed a network of quantum sensors embedded in superconducting materials. This sensor grid monitored background quantum fluctuations and automatically adjusted the facility's defensive systems to maintain optimal research conditions. The bedrock itself was chosen for its unusual mineral composition, which naturally screened out certain forms of quantum noise.

Central to the facility's design was the Quantum Core, a spherical chamber suspended within a vacuum-maintained space at the complex's heart. This chamber, constructed from exotic materials including artificially grown crystals, provided an environment where quantum effects could manifest with unprecedented clarity and control. The sphere's surface featured precisely arranged patterns that created standing waves of quantum energy.

Environmental control systems throughout Laboratory Seven maintained exact specifications of temperature, humidity, and electromagnetic shielding. Advanced air filtration removed all particles larger than a nanometer, while electromagnetic dampeners cancelled external radiation. The facility's power supply came from a dedicated fusion reactor buried 300 feet below ground, ensuring uninterrupted energy flow.

Dr. Elena Petrov supervised the construction of specialized research chambers that could be reconfigured for different experiments. These modular spaces featured walls lined with programmable metamaterials that could alter their quantum properties on demand. Ceiling-mounted arrays of quantum field generators could create precise experimental conditions anywhere within these chambers.

The facility's data center incorporated quantum-resistant shielding and utilized novel storage technologies that could handle the massive amounts of information generated by quantum experiments. Fiber optic networks with quantum encryption connected all sections of the facility, while a separate quantum communication system provided instantaneous data transfer between research areas.

Security measures at Laboratory Seven went beyond conventional approaches. Quantum entanglement-based detection systems monitored every cubic centimeter of space within and around the facility. These systems could detect any unauthorized presence or attempt to observe sensitive areas through quantum means. Access control utilized biometric scanners that checked quantum signatures unique to authorized personnel.

The facility's outer shell employed adaptive materials that responded to environmental conditions, automatically adjusting their properties to maintain internal stability. This living skin could heal minor damage and modify its structure to counter any threats to the quantum integrity of the research spaces. Solar collection panels integrated into the surface powered secondary systems while adding another layer of electromagnetic shielding.

Underground laboratories extended through three subterranean levels, each deeper level providing greater isolation from surface disturbances. The lowest level housed the most sensitive experiments, where quantum effects could be studied in near-perfect isolation. These chambers floated on quantum-stabilized magnetic fields that eliminated virtually all vibration.

Dr. Thompson's team designed specialized equipment storage areas with quantum-neutral zones that prevented unintended interactions between sensitive instruments. These storage spaces utilized controlled quantum fields to maintain equipment calibration and prevent degradation of crucial components.

The facility's ventilation system incorporated quantum filters that could remove unwanted particles while maintaining specific quantum states within research areas. Air circulation patterns were designed to create stable vortices that enhanced quantum coherence in experimental zones.

Construction crews worked under strict protocols to maintain quantum integrity during the building process. Each worker underwent quantum field training and used specially designed tools that minimized disruption to the facility's carefully planned quantum architecture. Assembly sequences followed precise timing patterns that harmonized with natural quantum frequencies.

Laboratory Seven's construction represented a breakthrough in quantum-aware architecture, demonstrating that buildings could actively participate in quantum research rather than merely containing it. The facility's success proved that understanding and working with quantum principles at every level of construction could create environments where the seemingly impossible became routine. This achievement set new standards for research facility design and construction worldwide.

Research Timeline

Laboratory Seven's groundbreaking discoveries unfolded across a meticulously documented timeline spanning fifteen years of intensive research. The initial phase began in 2010 when Dr. Elena Petrov assembled her core team of quantum physicists, mathematicians, and interdisciplinary specialists from leading institutions worldwide.

During the first year, researchers focused on establishing baseline measurements and calibrating the facility's advanced equipment. Dr. Thompson's preliminary experiments revealed unexpected quantum behaviors that would later become central to the laboratory's most significant discoveries. By early 2011, the team had documented the first instances of sustained quantum coherence at room temperature.

The breakthrough year of 2012 marked the discovery of quantum field resonance patterns that defied conventional physics models. Dr. Chen's group demonstrated repeatable experiments showing controlled quantum effects persisting in biological systems, leading to the establishment of the bio-quantum research division in late 2012.

2013 brought the first successful demonstrations of quantum-enhanced material properties. The research team documented cases where ordinary materials temporarily exhibited extraordinary characteristics when exposed to specific quantum fields. These findings laid the groundwork for future applications in materials science and engineering.

The period between 2014 and 2015 saw rapid advancement in understanding quantum-matter interactions. Dr. Morrison's team developed new theoretical frameworks to explain the observed phenomena, while practical applications began emerging from the laboratory's engineering division. By mid-2015, the first quantum-enhanced prototypes were undergoing preliminary testing.

2016 marked a pivotal year when researchers successfully maintained stable quantum states in complex systems for extended periods. This achievement led to the development of new experimental protocols that would become standard practice in quantum research facilities worldwide. The team's publications from this period revolutionized approaches to quantum experimentation.

During 2017, Laboratory Seven focused on exploring practical applications of their discoveries. Dr. Zhang's group demonstrated the first reliable quantum communication system operating outside controlled laboratory conditions. The year ended with successful tests of quantum-enhanced materials in industrial settings.

The research timeline reached another milestone in 2018 when the team documented unprecedented interactions between quantum fields and gravitational effects. These observations suggested new possibilities for understanding the relationship between quantum mechanics and gravity, leading to intense collaboration with theoretical physicists globally.

2019 brought breakthroughs in quantum computation using the laboratory's unique approach to maintaining quantum coherence. The research team demonstrated computing capabilities that surpassed traditional limitations, though the results remained difficult to replicate outside Laboratory Seven's specialized environment.

The year 2020 saw the successful integration of multiple research streams as Dr. Petrov's team combined insights from different experimental areas. This synthesis led to new understanding of quantum phenomena and their practical applications, resulting in several breakthrough patents.

In 2021, researchers achieved stable quantum entanglement across macroscopic distances without requiring extreme environmental conditions. This development opened new possibilities for quantum communication and sensing technologies, attracting significant attention from both scientific and industrial partners.

2022 marked the beginning of practical applications beyond the laboratory setting. The team successfully demonstrated quantum-enhanced materials in real-world conditions, while maintaining their unusual properties for commercially viable durations. Industrial partnerships began implementing these discoveries in prototype products.

The research timeline entered a new phase in 2023 as Laboratory Seven expanded its focus to include biological applications of quantum effects. Dr. Wong's team documented how living systems naturally utilized quantum processes, leading to new approaches in medical research and biotechnology.

By 2024, the laboratory had established itself as the world's leading center for quantum research application. The team's discoveries influenced fields ranging from materials science to medical technology, while continuing to push the boundaries of theoretical physics understanding.

The most recent period of 2025 has seen the integration of fourteen years of research into comprehensive theoretical frameworks that explain previously mysterious quantum phenomena. These new models have begun revolutionizing our understanding of fundamental physics while generating practical applications across multiple industries.

Throughout this timeline, Laboratory Seven maintained rigorous documentation and peer review processes, ensuring that each discovery was thoroughly verified before publication. The research program's systematic approach allowed for steady progress while maintaining scientific integrity and reproducibility of results.

The timeline reveals a carefully orchestrated progression from theoretical understanding to practical application, with each discovery building upon previous findings. This methodical approach has ensured that Laboratory Seven's breakthroughs remain both scientifically sound and commercially viable.

Looking forward, the research timeline continues to evolve as new discoveries emerge from ongoing experiments. The laboratory's commitment to long-term research goals, combined with its unique facilities and talented team, suggests that future years will bring even more remarkable advances in quantum science and its applications.

The Incident

Darkness descended upon Laboratory Seven at precisely 3:47 AM on October 12, 2024. The facility's quantum sensors registered an anomalous surge that defied all known physical laws, marking the beginning of what would become known simply as "The Incident."

Dr. Elena Petrov was working late in the Quantum Core chamber when the first signs appeared. The spherical chamber's monitoring systems detected unusual fluctuations in the quantum field, displaying patterns never before observed. Within seconds, the carefully maintained quantum stability began to unravel.

Emergency protocols activated automatically as the quantum disturbance spread through the facility's interconnected chambers. Dr. Thompson, monitoring the night shift from the control room, watched in disbelief as readings across all instruments spiked simultaneously. The quantum field generators, designed to maintain precise experimental conditions, began resonating at frequencies beyond their operational parameters.

Security systems registered multiple breaches in the quantum containment fields, yet no physical barriers had been compromised. Dr. Morrison's team, responding to the emergency alerts, discovered that space itself seemed to be folding within certain sections of the facility. Objects appeared to exist in multiple locations simultaneously, while time measurements became unreliable.

The incident escalated when the quantum core began emitting previously unknown forms of radiation. These emissions passed through the facility's sophisticated shielding as if it didn't exist, creating cascading effects throughout the laboratory's sensitive equipment. Dr. Chen, arriving on site within minutes of the first alert, documented reality-bending phenomena that challenged fundamental physics principles.

As the quantum disturbance intensified, researchers observed localized time dilation effects. Security footage revealed that events in different sections of the facility were occurring at different temporal rates. In some areas, seconds stretched into minutes, while in others, minutes compressed into moments.

The facility's quantum-aware architecture, designed to enhance research capabilities, began responding to the anomaly in unexpected ways. Walls shifted their molecular structure, attempting to contain the spreading quantum instability. The building's crystalline matrices, normally used to maintain quantum coherence, started exhibiting properties of conscious adaptation.

Dr. Wong's quantum detection systems recorded what appeared to be tears in the fabric of spacetime, though these ruptures remained mercifully small and contained within the facility's most heavily shielded areas. Through these tears, researchers glimpsed what seemed to be alternate versions of the laboratory, each showing slightly different configurations and events.

The crisis peaked when the quantum core's containment field began to collapse. Dr. Petrov, risking her own safety, initiated an experimental shutdown sequence that had never been tested under real conditions. The procedure required precise timing across multiple systems, complicated by the temporal distortions affecting different sections of the facility.

Laboratory Seven's advanced safety systems engaged in a complex dance of countermeasures, each adjustment triggering new quantum interactions that required immediate response. The facility's quantum-enhanced computers struggled to process the flood of contradictory data, as reality itself seemed to fluctuate between multiple states.

After twenty-seven minutes that simultaneously felt like hours and seconds, the quantum disturbance began to subside. The facility's systems gradually returned to normal operation, though evidence of the incident remained encoded in both instrumental records and the modified structure of affected materials.

In the aftermath, research teams discovered that certain sections of the laboratory had been permanently altered. Rooms that had experienced the strongest quantum fluctuations now exhibited unusual properties, including localized variations in fundamental constants and persistent quantum effects that defied explanation.

The incident's most profound impact came from the data collected during those twenty-seven minutes. Analysis revealed patterns suggesting that reality itself possessed more layers and flexibility than previously imagined. The quantum disturbance had temporarily lifted the veil between these layers, providing unprecedented insights into the nature of existence.

Dr. Petrov's team spent months analyzing the incident's causes and effects. Their findings indicated that Laboratory Seven had accidentally created conditions that amplified naturally occurring quantum fluctuations to macroscopic levels. This discovery led to new safety protocols and research directions, while also raising questions about the fundamental stability of spacetime itself.

The incident demonstrated both the risks and potential of advanced quantum research. While the facility's safety systems had prevented catastrophic failure, the event showed how much remained unknown about the quantum realm's interaction with classical reality. The experience provided valuable data for improving future containment systems and understanding the boundaries of quantum manipulation.

Years later, researchers still study the subtle changes left behind in Laboratory Seven's structure. These persistent anomalies serve as both warning and inspiration, reminding us that our exploration of quantum phenomena must proceed with caution while remaining open to unprecedented possibilities. The incident stands as a pivotal moment in scientific history, marking the point where humanity first glimpsed the true complexity of quantum reality.

Previous Experiments

Scientific breakthroughs rarely emerge from isolation, and Laboratory Seven's achievements stand upon the foundation of numerous groundbreaking experiments conducted across decades of quantum research. The earliest relevant experiments began in the 1970s when Dr. Heinrich Mueller demonstrated the first controlled quantum state manipulation at the University of Munich.

Mueller's work, though limited by the technology of his time, established crucial principles for maintaining quantum coherence. His team developed novel approaches to electromagnetic shielding that would later influence Laboratory Seven's design. The experiments, while primitive by modern standards, proved that quantum effects could be deliberately manipulated under laboratory conditions.

During the 1980s, Dr. Sarah Chen pioneered work on quantum field interactions at Berkeley. Her experiments revealed previously unknown relationships between quantum states and electromagnetic fields. Using custom-built equipment, Chen documented quantum behaviors that defied classical physics explanations, laying groundwork for future breakthroughs.

The Tokyo Institute of Technology contributed significantly through Dr. Hiroshi Tanaka's experiments in the early 1990s. His team achieved the first sustained quantum coherence in room-temperature conditions, though only for microseconds. These experiments provided crucial data about environmental factors affecting quantum states, informing later facility designs.

By the mid-1990s, Dr. Marcus Wong's research group at Cambridge had begun exploring quantum effects in crystalline structures. Their experiments demonstrated how certain crystal arrangements could naturally enhance quantum phenomena, leading to the development of quantum-aware architecture principles. Wong's findings directly influenced Laboratory Seven's structural design.

The late 1990s saw breakthrough experiments at CERN, where Dr. Elena Petrov first observed quantum field resonance patterns. Using advanced particle detectors, her team documented unexpected quantum behaviors that suggested the possibility of controlling quantum effects at macroscopic scales. These observations would later become central to Laboratory Seven's research program.

Russian scientists at the Moscow Institute of Physics contributed through experiments conducted between 2000 and 2005. Dr. Vladimir Korsakoff's team developed new methods for measuring quantum field interactions, creating many of the monitoring systems now standard in quantum research facilities. Their work established crucial safety protocols for high-energy quantum experiments.

Parallel research at MIT under Dr. James Thompson explored quantum-material interactions. His experiments revealed how different materials responded to quantum manipulation, leading to the development of specialized construction materials. Thompson's work provided essential data about containing and directing quantum effects within controlled environments.

The Chinese Academy of Sciences conducted crucial experiments in quantum communication during the early 2000s. Dr. Li Wei's team achieved quantum entanglement across unprecedented distances, demonstrating principles that would later enable Laboratory Seven's breakthrough in stable quantum state maintenance.

Between 2005 and 2010, Dr. Morrison's experiments at CalTech explored the relationship between quantum effects and gravitational fields. These studies revealed subtle interactions that influenced the design of Laboratory Seven's underground research chambers and led to new understanding of quantum field behavior.

Australian researchers at the Queensland Quantum Institute contributed through experiments in biological quantum effects. Dr. Amanda Foster's team documented how living systems naturally utilized quantum processes, opening new research directions that Laboratory Seven would later expand upon.

The Indian Institute of Technology's experiments under Dr. Raj Patel demonstrated quantum field manipulation techniques that became fundamental to Laboratory Seven's research methodology. Patel's team developed novel approaches to controlling quantum states while minimizing environmental interference.

European collaboration at CERN produced significant experiments in particle-quantum interactions between 2008 and 2010. These studies provided crucial data about high-energy quantum effects and established safety parameters for advanced quantum research facilities.

Canadian researchers at the University of British Columbia conducted experiments in quantum material science that directly influenced Laboratory Seven's construction. Dr. Marie Dumont's team developed new materials with enhanced quantum properties, several of which were incorporated into the facility's structure.

South Korean experiments at the Seoul National University explored quantum computing applications that would later complement Laboratory Seven's research. Dr. Park Sun-ji's team demonstrated novel approaches to maintaining quantum coherence in computing systems.

Brazilian scientists at the University of São Paulo contributed through experiments in quantum field detection. Dr. Carlos Santos developed sensitive measurement techniques that became essential to Laboratory Seven's monitoring systems.

These diverse experiments, conducted across multiple continents and decades, created the scientific foundation upon which Laboratory Seven built its revolutionary research program. Each contribution added crucial understanding about quantum phenomena, measurement techniques, safety requirements, and practical applications.

The collective knowledge gained from these previous experiments informed every aspect of Laboratory Seven's design and operation. From facility construction to experimental protocols, each element reflected lessons learned through years of international research effort. This accumulation of experience and data enabled Laboratory Seven to push beyond previous limitations and achieve unprecedented breakthroughs in quantum science.

Chapter 5: The Science Behind the Shadows

Quantum Entanglement

Quantum entanglement represents one of Laboratory Seven's most significant research areas, yielding discoveries that revolutionized our understanding of fundamental physics. The phenomenon, once described as "spooky action at a distance" by Einstein, became the cornerstone of numerous breakthrough experiments conducted within the facility's specialized chambers.

Dr. Elena Petrov's team first achieved sustained macroscopic entanglement in 2018, maintaining quantum correlation between objects visible to the naked eye. This achievement marked a departure from previous experiments that only demonstrated entanglement at the subatomic level. Using the facility's unique quantum field generators, researchers created stable entangled states between crystalline structures separated by several meters.

The breakthrough came through innovative application of quantum field resonance, a technique developed specifically within Laboratory Seven. By carefully manipulating overlapping quantum fields, researchers discovered methods to extend entanglement duration from microseconds to several minutes. This extended timeframe opened new possibilities for practical applications and detailed observation of entangled states.

Perhaps most remarkably, Dr. Thompson's group demonstrated that entanglement could be transferred between different types of quantum systems. They successfully linked quantum states in crystalline structures to those in specially prepared biological samples, suggesting that living systems might naturally utilize quantum entanglement processes.

The facility's quantum-aware architecture proved crucial in these experiments. Specialized chambers, designed to maintain precise quantum conditions, allowed researchers to study entanglement effects without environmental interference. The building's crystalline matrices actively supported entanglement experiments by providing stable quantum field environments.

Dr. Morrison's theoretical work provided crucial insights into the nature of entanglement itself. His mathematical models suggested that quantum correlations existed within a deeper layer of reality, one that transcended conventional spacetime limitations. These theories found experimental support when Laboratory Seven documented instantaneous state changes between entangled particles separated by significant distances.

The research team made an unexpected discovery when they observed that certain materials, when exposed to controlled quantum fields, developed spontaneous entanglement patterns. This phenomenon, dubbed "quantum crystallization," suggested that entanglement might be a more fundamental property of matter than previously understood.

Through careful experimentation, researchers identified specific conditions that enhanced entanglement stability. They found that certain crystalline structures, when properly aligned with quantum fields, could maintain entangled states for extended periods without active intervention. This discovery led to the development of quantum-enhanced materials with practical applications in communications and computing.

Dr. Chen's group demonstrated that entangled states could be manipulated with unprecedented precision using Laboratory Seven's advanced quantum field generators. They developed techniques to create complex networks of entangled particles, effectively building quantum circuits that operated through entanglement rather than traditional electronic connections.

The facility's unique capabilities allowed researchers to study how gravitational effects influenced quantum entanglement. They discovered subtle interactions between gravity and entangled states, providing new insights into the relationship between quantum mechanics and general relativity. These findings suggested possible approaches to unified field theory.

Laboratory Seven's experiments revealed that entanglement exhibited hierarchical properties, with quantum correlations forming nested patterns across different scales. This discovery led to new understanding of how quantum effects might influence macroscopic systems, challenging traditional boundaries between quantum and classical physics.

Researchers documented unexpected coherence effects when multiple entangled systems interacted. These interactions produced emergent properties that couldn't be predicted from individual entangled pairs, suggesting new possibilities for quantum computation and communication technologies.

The team developed novel methods for measuring entanglement strength and stability, creating standardized protocols that have since been adopted by research facilities worldwide. These measurement techniques proved essential for reproducible experiments and practical applications of quantum entanglement.

Through systematic investigation, Laboratory Seven identified specific frequency patterns in quantum fields that enhanced entanglement stability. These "resonant frequencies" became crucial tools for maintaining and manipulating entangled states, leading to more reliable quantum experiments.

Dr. Wong's research group demonstrated that entangled states could be transferred between different physical systems while maintaining coherence. This achievement opened new possibilities for quantum information storage and processing, suggesting ways to create hybrid quantum systems combining different technologies.

The facility's experiments showed that entanglement could be used to create quantum sensors with unprecedented sensitivity. By utilizing entangled particles as probes, researchers developed methods to detect subtle changes in physical fields and forces that were previously unmeasurable.

These groundbreaking discoveries continue to influence quantum research worldwide, establishing Laboratory Seven as a leading center for entanglement studies. The facility's unique combination of advanced technology, theoretical insight, and experimental expertise has pushed the boundaries of what's possible in quantum science, while opening new frontiers in our understanding of reality's fundamental nature.

Dimensional Overlap

The first signs of dimensional overlap appeared during a routine quantum field experiment in Laboratory Seven's Beta Chamber.

Dr. Sarah Chen noticed unusual patterns in the quantum resonance readings that didn't match any known phenomena. The instruments detected what appeared to be shadows of alternative quantum states, suggesting the presence of parallel dimensional structures intersecting with our own.

Through careful manipulation of the quantum fields, researchers discovered they could enhance these dimensional intersections, making them more pronounced and stable. The overlapping regions exhibited unique properties, including localized variations in fundamental physical constants and unexpected quantum behaviors. Within these zones, particles seemed to exist in multiple states simultaneously, beyond normal quantum superposition.

Dr. Thompson's team developed specialized sensors capable of mapping these dimensional interfaces. Their measurements revealed intricate geometric patterns in the overlap regions, suggesting an underlying mathematical structure to the phenomenon. These patterns appeared to follow complex fractals, with similar features repeating at different scales.

The breakthrough came when Dr. Morrison successfully established controlled dimensional resonance. By carefully tuning the facility's quantum field generators, his team created stable zones where multiple dimensional states coexisted. Within these regions, researchers observed phenomena that challenged conventional physics, including objects that appeared to exist in slightly different positions simultaneously.

Laboratory Seven's unique architecture proved crucial in maintaining these dimensional overlaps. The building's quantum-aware design helped stabilize the intersection points, preventing them from collapsing or expanding uncontrollably.

The crystalline matrices embedded in the facility's structure appeared to naturally enhance dimensional resonance, creating more stable and predictable overlap zones.

Dr. Petrov's research revealed that living organisms responded uniquely to dimensional overlap regions. When exposed to these zones, biological samples exhibited unexpected quantum coherence effects, suggesting that life might naturally interact with multiple dimensional states. This discovery opened new avenues for understanding consciousness and biological processes.

The team documented strange temporal effects within the overlap zones. Time appeared to flow at slightly different rates depending on the degree of dimensional intersection. These variations, while minimal, provided crucial insights into the relationship between time, space, and dimensional structure.

Through careful experimentation, researchers identified specific quantum field configurations that enhanced dimensional overlap stability. They discovered that certain frequencies created resonant patterns that strengthened the intersection points while minimizing unwanted effects. This knowledge proved essential for conducting controlled experiments within these unique regions.

Dr. Wong's group observed that materials exposed to dimensional overlap zones sometimes retained unusual properties even after removal. These altered substances exhibited quantum behaviors at macroscopic scales, suggesting possible applications in advanced materials science. The changes appeared permanent, indicating fundamental modifications to the materials' quantum structure.

The facility's advanced monitoring systems revealed subtle energy exchanges between overlapping dimensions. These transfers followed precise mathematical patterns, suggesting an underlying order to interdimensional interactions. The energy flows appeared to maintain dimensional stability while allowing limited information exchange between states.

Researchers discovered that dimensional overlaps could be used to enhance quantum entanglement effects. Particles entangled within these regions maintained stronger correlations and exhibited new properties not seen in normal space. This enhancement led to breakthroughs in quantum communication and computation research.

Dr. Chen's team developed methods to map the topology of dimensional intersections, creating detailed models of how different reality states interacted. Their work revealed complex networks of dimensional connections, suggesting that our universe might be part of a larger dimensional ecosystem.

The facility's experiments showed that gravitational effects became more complex within overlap zones. Gravity appeared to respond to mass distributions across multiple dimensional states, creating unusual force patterns that required new mathematical frameworks to describe.

Laboratory Seven's research indicated that dimensional overlaps might occur naturally at quantum scales, only becoming noticeable when amplified by specialized equipment. This discovery suggested that nature might regularly utilize dimensional interactions in ways not previously recognized by science.

Through systematic investigation, researchers identified specific conditions that promoted or inhibited dimensional intersection. They learned to create precisely controlled overlap regions for experimental purposes while developing safety protocols to prevent unintended dimensional interactions.

The study of dimensional overlap has revolutionized our understanding of reality's structure. What began as unexpected readings in a quantum experiment has evolved into a new field of physics, promising applications in energy production, materials science, and communication technology. Laboratory Seven's pioneering work continues to expand our knowledge of dimensional phenomena, while raising fascinating questions about the nature of existence itself.

Chapter 6: Human Interaction

Researcher Experiences

The corridors of Laboratory Seven echo with the personal stories of dedicated scientists who've witnessed phenomena that challenge our understanding of reality. Dr. Elena Petrov recalls her first encounter with quantum field resonance as a moment that fundamentally changed her perspective on physics. "The instruments showed readings that shouldn't have been possible," she notes in her research journal. "Yet there they were, defying everything we thought we knew."

Dr. James Thompson describes the intense preparation required for each experiment. "We spent months calibrating equipment and running simulations before attempting anything new," he explains. The team's commitment to precision became legendary, with researchers often working through the night to maintain optimal conditions during critical experiments.

Sarah Chen remembers the exact moment when she first observed dimensional overlap effects. "The air seemed to shimmer, like heat waves rising from hot pavement, but in a geometrically precise pattern," she writes. Her detailed observations helped establish the initial protocols for studying these phenomena safely.

The facility's unique environment affected researchers in unexpected ways. Dr. Morrison reports experiencing unusual sensory phenomena during extended work sessions in the quantum chambers. "Sometimes I felt as if I could perceive the quantum fields directly," he notes, "though colleagues suggest this might be an effect of prolonged exposure to the specialized environment."

Junior researcher Maria Rodriguez documented the emotional impact of witnessing quantum entanglement at macroscopic scales. "Seeing two objects maintain perfect correlation despite their separation challenged my fundamental assumptions about reality," she reflects. Her experiences mirror those of many team members who found their scientific worldview transformed by their work.

The demanding nature of the research created strong bonds among team members. Dr. Wong describes how shared challenges and breakthroughs fostered a unique camaraderie. "We celebrated each success and supported each other through failed experiments," he recalls. "The intensity of our work created friendships that extend beyond professional relationships."

Laboratory Seven's security protocols initially felt restrictive to many researchers. Dr. Tanaka admits feeling frustrated by the extensive clearance procedures but later came to appreciate their necessity. "Once we understood the full implications of our discoveries, the precautions made perfect sense," he reflects in his personal notes.

The facility's isolation posed challenges for maintaining work-life balance. Dr. Foster describes developing new routines to cope with extended research periods. "We learned to create our own community within the facility," she explains. "Regular social gatherings became crucial for maintaining morale and mental well-being."

Unexpected observations often led to new research directions. Dr. Korsakov recounts how a seemingly failed experiment revealed previously unknown quantum effects. "What we initially considered experimental error turned out to be our most significant discovery," he notes, emphasizing the importance of remaining open to unexpected results.

The physical demands of working with sensitive equipment required researchers to develop new skills. Dr. Li describes learning to move with extraordinary precision to avoid disturbing delicate experimental setups. "Even our breathing patterns had to be controlled during certain procedures," she explains.

Personal protective equipment became a constant companion for researchers working in high-energy areas. Dr. Patel recalls the strange sensation of conducting experiments while wearing specialized suits designed to shield against quantum field effects. "It felt like being an astronaut exploring an alien environment," he writes.

The facility's unique architecture affected researchers' perceptions of space and time. Dr. Thompson notes how the quantum-aware design created subtle psychological effects. "The building seemed to have its own consciousness," he observes, "responding to our presence in ways we're still trying to understand."

Many researchers report experiencing profound philosophical shifts during their time at Laboratory Seven. Dr. Chen describes questioning fundamental assumptions about reality. "Our work here suggests that the universe is far more complex and interconnected than we previously imagined," she reflects.

The pressure of working with groundbreaking technology took its toll. Dr. Morrison acknowledges the stress of handling potentially revolutionary discoveries. "We were acutely aware that our findings could fundamentally change human understanding of physics," he notes.

Despite challenges, the sense of purpose remained strong among team members. Dr. Rodriguez expresses the common sentiment: "Every day brought the possibility of discovering something that no one had ever seen before." This shared excitement helped sustain researchers through difficult periods.

The experience of working at Laboratory Seven has permanently altered how these scientists view their field. Dr. Petrov summarizes the transformation: "We came here as conventional physicists and became explorers of reality's deepest mysteries." Their personal accounts provide crucial insight into the human experience of pushing the boundaries of scientific knowledge.

Communication Attempts

Laboratory Seven's first successful communication attempt occurred during a routine quantum entanglement experiment. Dr. Elena Petrov noticed unusual patterns in the quantum noise that couldn't be attributed to normal background interference. These patterns exhibited a structured regularity that suggested intentional manipulation of quantum states.

The breakthrough came when Dr. Thompson's team developed specialized quantum field sensors capable of detecting subtle variations in dimensional overlap zones. These instruments revealed complex sequences of quantum state changes that appeared to carry encoded information. The patterns were too precise and regular to be random fluctuations, suggesting an external intelligence was attempting to establish contact.

Initial efforts to decode these signals proved challenging. Dr. Morrison developed a mathematical framework to analyze the quantum patterns, revealing underlying structures that resembled known communication protocols. The team observed that these signals often appeared strongest during periods of enhanced dimensional overlap, suggesting a connection between different reality states.

The facility's quantum-aware architecture played a crucial role in receiving and maintaining these communications. The building's crystalline matrices appeared to naturally amplify certain quantum frequencies, making it easier to detect and study the incoming signals. Researchers discovered that specific areas within Laboratory Seven acted as natural focal points for these transmissions.

Dr. Chen's group made significant progress in understanding the communication method when they identified recurring patterns that resembled mathematical constants. These universal values appeared encoded within the quantum signals, suggesting an attempt to establish a common language based on fundamental physical principles.

The team developed new protocols for responding to these signals. Using precisely controlled quantum field generators, researchers created patterns that mirrored the received communications. They observed that these responses triggered changes in the incoming signals, suggesting successful two-way interaction.

Laboratory Seven's specialized equipment allowed researchers to map the spatial distribution of these communications. They discovered that the signals appeared to originate from multiple sources, each with distinct characteristics. This discovery suggested the possibility of contact with various dimensional entities or civilizations.

Dr. Wong's research revealed that biological systems showed unusual responses during periods of active communication. Living tissue exhibited enhanced quantum coherence effects, suggesting that organic matter might naturally resonate with these quantum signals. This discovery opened new avenues for understanding potential biological interfaces for interdimensional communication.

The facility's experiments demonstrated that temporal factors influenced communication clarity. Signals appeared stronger during specific astronomical alignments and quantum field conditions. Researchers developed predictive models to identify optimal windows for communication attempts.

Through careful analysis, the team identified several distinct "languages" within the quantum signals. Each exhibited unique mathematical structures and encoding methods, suggesting different origins or purposes. Dr. Petrov's group focused on decoding what appeared to be the simplest of these patterns, making steady progress in understanding their basic elements.

Laboratory Seven's research indicated that successful communication required precise control of quantum field parameters. Minor variations in field strength or frequency could disrupt the signal reception, making consistency crucial for maintaining contact. The team developed automated systems to maintain optimal conditions during communication attempts.

Dr. Morrison's theoretical work suggested that these communications might represent natural phenomena rather than artificial signals. However, the precise mathematical nature of the patterns strongly indicated deliberate intention behind their creation. This apparent paradox led to new theories about the relationship between consciousness and quantum mechanics.

The facility's advanced monitoring systems revealed that successful communication attempts created subtle changes in local space-time properties. These effects were minimal but measurable, suggesting that information exchange between dimensions involved actual physical interactions rather than purely quantum phenomena.

Researchers discovered that certain materials enhanced communication clarity when placed within quantum field generators. These substances, primarily crystalline structures with specific atomic arrangements, appeared to act as natural amplifiers for the quantum signals. This finding led to the development of more effective communication equipment.

Dr. Chen's team observed that communication attempts sometimes triggered unexpected dimensional overlap effects. These incidents, while initially concerning, provided valuable data about the relationship between quantum communication and dimensional boundaries. New safety protocols were established to manage these interactions effectively.

The study of these communication attempts has revolutionized our understanding of quantum information exchange and dimensional interactions. While many questions remain unanswered, Laboratory Seven's ongoing research continues to yield new insights into the nature of consciousness, reality, and the possibility of contact with other forms of intelligence. The team remains committed to developing more effective methods for establishing and maintaining these mysterious quantum communications.

Psychological Impact

Strange dreams began plaguing the research team three months into the quantum field experiments. Dr. Sarah Chen first documented these occurrences in her personal journal, describing vivid visions of geometric patterns that seemed to extend beyond three-dimensional space. Within weeks, similar reports emerged from other team members, suggesting a correlation between exposure to quantum field experiments and altered psychological states.

The facility's resident psychologist, Dr. Marcus Rivera, established a comprehensive monitoring program to track these psychological changes. Researchers reported experiencing heightened spatial awareness, unusual time perception, and recurring sensations of existing in multiple places simultaneously. These effects persisted even when team members were away from the experimental chambers, indicating lasting neurological changes.

Sleep patterns among the staff shifted dramatically. Dr. Thompson's monitoring devices recorded increased rapid eye movement activity and complex brainwave patterns previously unseen in sleep studies. Many researchers described dreams where they could perceive quantum phenomena directly, as if their minds had developed new sensory capabilities.

The isolation of Laboratory Seven intensified these psychological effects. The facility's remote location and strict security protocols created an environment where researchers became increasingly attuned to subtle quantum fluctuations. Dr. Petrov noted that long-term exposure to dimensional overlap zones appeared to enhance intuitive understanding of quantum mechanics among team members.

Stress manifested in unique ways within the facility. Rather than traditional anxiety symptoms, researchers experienced what Dr. Rivera termed "quantum anxiety" - a persistent sensation of reality's malleability. Some staff members reported difficulty maintaining conventional perspectives on causality and linear time, requiring specialized counseling sessions to manage these perceptual shifts.

Social dynamics within the facility evolved as shared experiences created deep bonds among team members. Researchers developed their own vocabulary to describe quantum phenomena, speaking in terms that outsiders found increasingly difficult to comprehend. This linguistic evolution reflected fundamental changes in how the team perceived and processed reality.

Dr. Morrison documented cases where researchers demonstrated unexpected abilities to predict quantum measurement outcomes. While initially dismissed as coincidence, statistical analysis revealed accuracy rates significantly above chance. These developments raised questions about human consciousness's potential interaction with quantum systems.

The facility's architecture began affecting staff psychology in subtle ways. Researchers reported feeling more connected to the building's quantum-aware design, describing it as an extension of their consciousness. Dr. Wong's team discovered that certain areas of the facility induced specific emotional and cognitive states, possibly due to localized quantum field effects.

Memory formation and recall showed interesting modifications. Team members experienced what they described as "quantum memories" - recollections that seemed to exist in superposition, containing multiple versions of events simultaneously. Dr. Rivera's studies suggested these effects might represent genuine alterations in how the brain processes information.

Physical symptoms accompanied psychological changes. Researchers reported enhanced sensory perception, particularly in areas related to pattern recognition and spatial awareness. Medical examinations revealed subtle changes in neural activity, suggesting adaptation to quantum field exposure.

The team's perception of time became increasingly fluid. Dr. Chen documented cases where researchers accurately sensed the timing of quantum events without instrumental assistance. These temporal anomalies appeared to strengthen during periods of intense dimensional overlap.

Creativity among staff members surged, particularly in mathematical and theoretical thinking. Researchers developed novel approaches to quantum mechanics that seemed to arise from intuitive understanding rather than conventional analysis. These insights often proved valuable in advancing the facility's research goals.

Family relationships faced unique challenges as researchers struggled to communicate their experiences to outside contacts. The psychological changes experienced within Laboratory Seven created an experiential gap that proved difficult to bridge. Support groups formed among family members to help cope with these challenges.

Dr. Rivera identified several stages of psychological adaptation to quantum exposure. Initial disorientation gave way to acceptance and eventually to integration of new perceptual abilities. This process varied among individuals but followed consistent patterns that helped in developing support protocols.

The long-term implications of these psychological changes remain under study. While some effects appear permanent, researchers have developed coping strategies to maintain functional connections with conventional reality. The team's experiences suggest human consciousness possesses greater flexibility and potential for adaptation than previously understood.

Regular psychological evaluations became essential for monitoring staff welfare. Dr. Rivera's team developed specialized techniques for helping researchers navigate their evolving mental states while maintaining scientific objectivity. These protocols now serve as the foundation for managing psychological health in quantum research facilities.

The psychological impact of quantum field exposure has revealed unexpected aspects of human consciousness and its relationship to reality. As Laboratory Seven continues its groundbreaking research, understanding and managing these mental transformations remains crucial for both scientific progress and personnel wellbeing.

Chapter 7: Implications and Theories

Scientific Applications

Quantum field manipulation techniques developed at Laboratory Seven have revolutionized multiple scientific disciplines. The breakthrough came when Dr. Elena Petrov discovered methods to stabilize dimensional overlap zones for extended periods. This achievement opened new avenues for research across physics, chemistry, biology, and materials science.

Medical applications emerged as an early success story. Dr. Thompson's team developed quantum field generators that could influence cellular repair mechanisms. By carefully controlling dimensional resonance patterns, researchers demonstrated accelerated healing in test tissues. These findings led to experimental treatments for previously untreatable conditions, though strict protocols limit their current application.

Materials science underwent dramatic advancement through quantum field research. Scientists discovered that exposure to controlled dimensional overlap could alter fundamental material properties. Dr. Wong's laboratory successfully created new crystalline structures with unprecedented strength-to-weight ratios. These materials exhibit unique quantum properties that make them valuable for specialized applications.

Biological research benefited from new insights into quantum effects in living systems. Dr. Chen's studies revealed that cellular processes utilize quantum mechanisms more extensively than previously thought. This understanding led to breakthroughs in biotechnology, including more efficient photosynthesis systems and enhanced protein folding techniques.

Energy generation systems evolved through the application of quantum field technology. Researchers developed methods to harvest energy from dimensional boundary interactions, creating highly efficient power sources. While currently limited to laboratory settings, these systems show promise for future clean energy production.

Chemical synthesis achieved new levels of precision through quantum field control. Scientists can now manipulate molecular interactions with unprecedented accuracy, leading to the development of novel compounds. These techniques have particular value in pharmaceutical research, enabling the creation of more effective drug delivery systems.

Environmental science applications emerged from quantum field studies. Researchers discovered methods to detect and influence environmental systems at the quantum level. These techniques show promise for pollution remediation and climate monitoring, though large-scale implementation remains challenging.

Communications technology advanced significantly through quantum field research. Dr. Morrison's team developed systems that utilize dimensional overlap effects for data transmission. These methods offer potential for secure, instantaneous communication, though current applications remain limited to laboratory conditions.

Agricultural applications demonstrated surprising potential. Quantum field exposure under controlled conditions showed positive effects on plant growth and resistance to disease. These findings suggest possibilities for improving crop yields and developing more resilient food sources.

Geological studies benefited from new quantum sensing technologies. Researchers developed methods to detect subsurface structures and composition through quantum field interactions. These techniques provide more detailed information than conventional surveying methods.

Space research applications emerged as scientists better understood quantum field effects. Laboratory Seven's discoveries suggest new approaches to propulsion and navigation systems. These developments could revolutionize space exploration, though practical implementation requires further research.

Neuroscience advanced through studies of quantum effects in brain function. Researchers identified previously unknown mechanisms of neural communication that operate at the quantum level. These findings offer new approaches to treating neurological conditions and understanding consciousness.

Manufacturing processes improved through quantum field applications. Scientists developed methods for precise material manipulation at the atomic level, enabling new manufacturing techniques. These advances show particular promise for nanotechnology applications.

Water purification systems benefited from quantum field technology. Researchers discovered methods to remove contaminants through dimensional resonance effects. These techniques offer more efficient and environmentally friendly approaches to water treatment.

Climate science gained new tools through quantum field research. Scientists developed more accurate methods for modeling complex environmental systems using quantum computing techniques enhanced by dimensional overlap effects. These models provide better predictions of climate patterns and changes.

Transportation systems could be transformed by quantum field applications. Research suggests possibilities for new propulsion methods and energy storage systems. While currently theoretical, these concepts show promise for future development.

Waste management improved through quantum field processing techniques. Researchers developed methods to break down complex materials using dimensional resonance effects. These approaches offer more efficient and environmentally sound waste treatment options.

Construction technology advanced through the application of quantum-enhanced materials. Scientists created building materials with superior properties by controlling quantum effects during formation. These materials offer improved strength, durability, and energy efficiency.

Security systems evolved through quantum field technology applications. Researchers developed new methods for detecting and preventing unauthorized access using quantum sensing techniques. These systems provide enhanced security while reducing false alarms.

The scientific applications of quantum field research continue to expand as Laboratory Seven makes new discoveries. Each breakthrough reveals additional possibilities for practical implementation across diverse fields. While many applications remain in early stages, the potential for transformative impact on science and technology grows increasingly clear.

Parallel Universe Evidence

Evidence of parallel universes first manifested during a routine quantum field experiment at Laboratory Seven when Dr. Thompson detected unexplained energy signatures. These signatures exhibited patterns that couldn't be explained by known physical phenomena, displaying characteristics that suggested interactions with alternate realities.

The breakthrough occurred when Dr. Chen's team captured high-resolution quantum field measurements showing distinct ripples in space-time fabric. These ripples appeared to correspond with moments of dimensional overlap, providing tangible evidence of parallel universe interactions. The patterns revealed structured variations that indicated organized systems existing alongside our own reality.

Physical artifacts began materializing within the facility's specialized containment chambers. These objects possessed properties that defied our universe's natural laws, exhibiting molecular structures impossible under normal conditions. Dr. Morrison's analysis confirmed these items originated from environments with different physical constants.

Electromagnetic anomalies provided further evidence of parallel universe interaction. Researchers documented consistent patterns of energy fluctuations that corresponded with periods of increased dimensional overlap. These fluctuations carried signatures of alternate physical laws, suggesting the temporary merging of different universal constants.

Dr. Petrov's team recorded instances of temporal inconsistencies within the facility. Local time measurements showed variations that indicated the influence of parallel timelines. These temporal anomalies became more pronounced during periods of strong dimensional resonance, providing measurable evidence of multiple time streams.

Quantum entanglement experiments revealed unprecedented results when conducted near dimensional overlap zones. Particles demonstrated coordination patterns that suggested interaction with counterparts in parallel universes. These findings supported theoretical models predicting the existence of quantum connections between different realities.

Biological samples exposed to overlap zones exhibited remarkable changes. Cell cultures developed characteristics that appeared adapted to different environmental conditions, suggesting influence from parallel evolution paths. Dr. Wong's research documented these adaptations, providing biological evidence of alternate reality exposure.

Gravitational measurements within Laboratory Seven showed peculiar variations. Sensors detected localized changes in gravitational fields that indicated the presence of mass and energy from other universes. These measurements helped map the boundaries between different reality states.

Chemical reactions near overlap zones produced unexpected compounds. These substances displayed properties that could only exist under different physical laws, offering tangible proof of parallel universe interaction. Some compounds remained stable even after the dimensional overlap subsided, allowing for detailed analysis.

Advanced imaging systems captured visual evidence of parallel universe phenomena. Specialized cameras documented brief glimpses of alternate spatial configurations, revealing environments that followed different geometric principles. These images provided crucial data for understanding parallel universe structures.

Energy readings throughout the facility showed patterns of interference from parallel power sources. These interactions created measurable fluctuations in local energy fields, suggesting the temporary merging of different universal energy systems. Dr. Thompson's team developed new instruments to track these energy exchanges.

Mathematical models based on collected data confirmed the theoretical predictions of parallel universe existence. The equations revealed underlying patterns that could only be explained by the presence of multiple realities operating simultaneously. These models helped researchers predict and measure dimensional overlap events.

Sound recordings near overlap zones captured acoustic phenomena from parallel universes. These recordings contained frequencies and patterns that couldn't originate in our reality, providing auditory evidence of alternate physical laws. Analysis revealed structured patterns suggesting intelligent organization.

Weather patterns around Laboratory Seven showed unusual variations during strong dimensional overlap events. Local atmospheric conditions exhibited characteristics that indicated influence from parallel climate systems. These meteorological anomalies provided additional evidence of reality intersection.

Radiation signatures unique to parallel universe interaction were documented throughout the facility. These emissions displayed properties inconsistent with known radioactive processes, suggesting origins in universes with different nuclear physics. Careful monitoring of these signatures helped track dimensional stability.

Material analysis of parallel universe artifacts revealed atomic structures impossible in our reality. These materials demonstrated properties that could only exist under different physical constants, providing tangible evidence of alternate universal laws. Some samples retained their unusual properties indefinitely.

Computer systems within the facility recorded data patterns indicating parallel information processing. These patterns suggested the temporary intersection of different computational realities, offering evidence of alternate technological development paths. The data provided insights into parallel evolution of information systems.

Geological samples collected near overlap zones showed evidence of alternate Earth histories. These specimens contained minerals and formations that could only form under different planetary conditions, suggesting parallel geological development. Analysis revealed consistent patterns indicating organized parallel realities.

The accumulating evidence at Laboratory Seven continues to support the existence of parallel universes. Each new discovery provides additional confirmation of multiple reality states, while raising fascinating questions about the nature of existence and consciousness. The research team remains dedicated to documenting and understanding these remarkable phenomena.

www.ingramcontent.com/pod-product-compliance
Lightning Source LLC
LaVergne TN
LVHW022310200926
848481LV00004B/1273
* 9 7 9 8 3 4 8 2 3 3 5 2 5 *